DODD, MEAD WONDER BOOKS

To Georgio Santelli

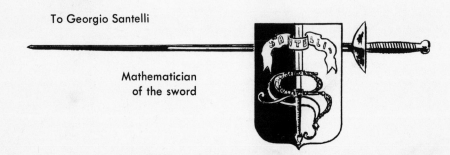

Mathematician
of the sword

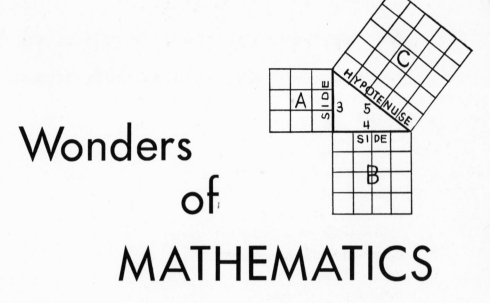

Wonders
of
MATHEMATICS

By Rocco Feravolo

ILLUSTRATED BY ROBERT BARTRAM

DODD, MEAD & COMPANY, NEW YORK

| 1 | 2 | 3 | 4 | 5 | 6 | 7 | 8 | 9 | 10 |

𒐕 𒐖 𒐗 𒐘 𒐙 𒐚 𒐛 𒐜 𒐝 𒌋

CONTENTS

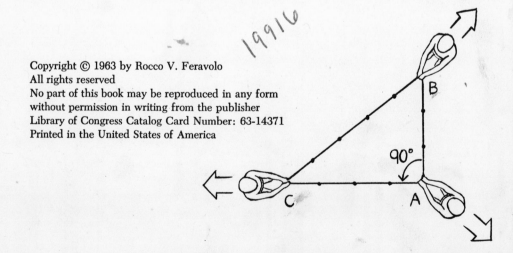

Long ago, man lived in caves. He did not have the knowledge or the tools to build a home out of wood or stone, so he had to be satisfied with what Mother Nature had provided for him. There were none of the modern conveniences we have today.

The only luxury our early ancestors enjoyed was fire. It took man many years to learn how to make a fire, but once he learned how, it was his most prized possession.

During the long evenings in the cave, the women of the household usually spent their time making clothes out of the furs of animals the men killed in the hunts. The men would occasionally spend their time drawing pictures on the walls of the cave by the firelight. Perhaps they were not great artists, but their drawings revealed many interesting stories about their lives and the hardships they had to endure.

Some of these cave drawings showed hunters throwing crude spears at the wild animals. In one of the drawings, five spears are in the body of an animal. The prehistoric artist who drew this picture did not know the *number* of times the hunters had hit the animal. There was no number system such as ours during those times. The number five was not known. The only way that early man could express numbers of things was in a "one-to-one" relationship. One mark or one symbol stood for one thing.

If a man saw three animals in the woods and wanted to record what he had seen, he would either scratch three lines in the dirt with a twig or place three rocks in a pile. He had no number words in his vocabulary to tell his friends what he had seen. Instead, he would show them the scratches in the dirt or the pile of rocks. This method of recording number situations was used by the men who tended the herds of animals.

For each animal that went to pasture daily, the keeper of the herd would place a rock on a pile. When dusk came, he would remove one rock at a time for each animal that returned. If there was a rock left in the pile, then he knew that one of the animals was missing.

One day, probably quite by accident, man realized that he could use parts of his body to remember a number of objects that he had seen. Instead of piling rocks or making marks on the ground, he used his fingers to tell his friends about the number of things that he saw.

If the number of objects was greater than ten, the cave man was baffled, because his finger system only went up to ten.

Then one day a cave man discovered he could also use his toes for recording objects. He may have been cooling his feet in a stream when he saw a large number of big fish. This was certainly something to tell his friends. But there were more fish than he had fingers. Did an insect perhaps land on his toe to make him think of using his toes to record the rest of the fish? We cannot be sure, because there is no written record of this fish story. But this may have been the way a number system based on twenty originated.

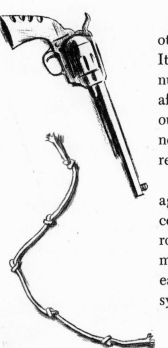

Notching a stick with a knife was another way early man kept track of things. It was used for many centuries before number systems were invented, and long afterward. Even though gun fighters in our Wild West knew how to count, they notched the handles of their six-shooters to record the number of their victims.

Rope was also used to keep records long ago. If a person wanted to remember a certain thing, he would make a knot in a rope. There existed in the course of time many number systems. But it took our early ancestors a long time to develop a system of writing numbers.

7

The oldest written record of mathematics was found in Egypt in the middle of the nineteenth century. It was called the Ahmes Papyrus. Experts say that it was written by an Egyptian scribe by the name of Ahmes. He copied it from the works of an unknown person. This great mathematical discovery was brought to England by A. Henry Rhind. Because of this, it is often referred to as the Rhind Papyrus.

The Ahmes Papyrus revealed to the world the many ways the early Egyptians used mathematics. They could add and subtract whole numbers, and had a knowledge of fractions, or parts of numbers. Of course, they used a different set of symbols for their numbers than we use today.

There were problems in the Ahmes Papyrus which showed that the Egyptians used a rough kind of trigonometry to make the faces of their pyramids slope at a desired angle. Trigonometry is a branch of mathematics which deals with the relationship of the sides and the angles of triangles.

Problems involving geometry were found in the Ahmes Papyrus, too. Geometry is that branch of mathematics which deals with the properties of solids, surfaces, lines, and angles. The Egyptians used geometry in their building projects and in surveying. They also used geometric designs as decorations.

Numbers are everywhere. They are used by all of us every day. No matter where you go, you will see numbers. They are found on clocks, books, machines, and on countless other things. Make a note of all the things you see in one day that use numbers. Do not be surprised if your list goes over a hundred. Compare your list with a friend's list. Are they alike?

Can you imagine a world without numbers? Write a story about "A Day in the World Without Numbers." It could start with you waking up and looking at a clock with no numbers on the dial.

We take numbers for granted. We add, divide, subtract, and multiply with them. The people of long ago had number systems, too, but they did not work like ours.

EGYPTIAN NUMBERS

The Egyptians were one of the first civilizations to develop a number system. They used hieroglyphic numerals. The number 1 was represented by a vertical line. The number 10 looked like an upside-down U. The number 100 looked something like the number 9 in our number system.

The number 1,000 was the picture of the lotus plant. The Egyptians even had a symbol for a million. It was the drawing of a man with raised arms. Perhaps he was showing amazement at such a large number.

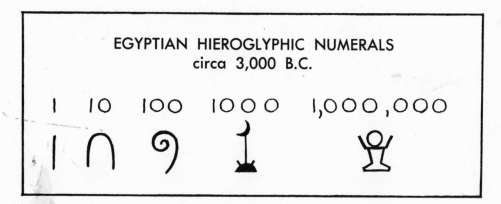

EGYPTIAN HIEROGLYPHIC NUMERALS
circa 3,000 B.C.

1   10   100   1000   1,000,000

The Egyptians wrote their numbers from left to right most of the time. They also wrote them vertically, as well as from right to left.

The number 3,401 can be written in the following ways using Egyptian numerals, yet each way means the same amount.

RIGHT TO LEFT	LEFT TO RIGHT	VERTICALLY

The Egyptians could write the same number in three different ways.

If we wrote the number 3,401 from right to left using our number system, the number would be 1,043. We cannot write our numbers from right to left without changing their values because our number system has a "place value." *Where* a figure is placed in a number makes a difference in its value. The number 3,401 is very different from 1,043 or 1,403 or 3,041 or 4,013.

ACTIVITY

Write the following numbers using Egyptian hieroglyphics. Write them from left to right.

303	211	301
1,396	879	650
21	37	89
41	32	5,637

(answers on page 63)

BABYLONIAN NUMBERS

The Babylonians made many contributions to the development of mathematics. They had a number system that used the bases of ten and sixty. The Babylonian merchants who lived about 2,000 B.C. did most of their arithmetic problems on clay tablets. They used a stylus to make impressions in the soft clay. The stylus made a wedge-shaped impression. The number on the clay tablet in this illustration is equal to 212.

Wedge-shaped impressions on a clay tablet

Each impression on the left is equal to sixty. The three smaller impressions in the middle each represent ten. Each impression on the right is equal to one. The number totals 60 + 60 + 60 + 10 + 10 + 10 + 1 + 1 or 212. The vertical impressions could either equal sixty or one in the Babylonian system. If they were to the left of the tens symbol, they represented sixty. If the impression was to the right of the tens symbol, the mark was equal to one. The Babylonian numbers cannot be written from right to left because of this reason. The Babylonians were the first people of the ancient world to use this place value idea. Where a symbol was placed determined its value.

EGYPTIAN

∩∩∩∩
∩∩∩∩ || = 82

LEFT TO RIGHT

|| ∩∩∩∩
∩∩∩∩ = 82

RIGHT TO LEFT

BABYLONIAN

Y ≼ YY = 82

YY ≼ Y = 141, NOT 82

Many tablets were found that showed how the Babylonians used simple arithmetic in their business records. This was the first kind of bookkeeping known to man. Today, electrical machines do the bookkeeping and many other jobs that are related to business. In fact, a business machine could, in less than an hour, do all the bookkeeping that was ever done by the early Babylonians.

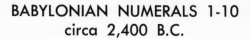

BABYLONIAN NUMERALS 1-10
circa 2,400 B.C.

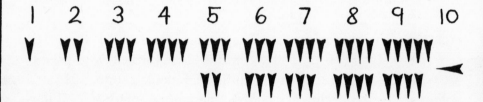

1 2 3 4 5 6 7 8 9 10

ACTIVITY

Make a clay tablet and write on it the way the early Babylonians did. You will need the following materials: a pencil and clay that will harden. Form the clay so that it looks like a Babylonian tablet. Make it about four inches square. Use a pencil to make the impressions in the tablet. Write the number 334.

Your tablet should look like the one in the illustration at the left.

Let the tablet dry. When it is completely dry, paint or shellac it. You can use it as a paper weight or mount it on the wall in your room.

The Babylonian place value idea did not include a zero to help locate the position of figures in a number. When only two vertical impressions were made, there was no way of knowing whether they were each equal to one or each equal to sixty.

GREEK NUMBER SYSTEMS

The Greeks had more than one number system. The oldest system used the twenty-four letters of the Greek alphabet. Many years later, the Greeks added three more letters to their number system. The twenty-seven letters represented the Ionic number system. The letters were divided into three groups. The first nine letters represented the numbers from 1 to 9.

1	2	3	4	5	6	7	8	9
A	B	Γ	Δ	E	F	Z	H	θ

The second group of letters represented the numbers from 10 to 90.

10	20	30	40	50	60	70	80	90
I	K	Λ	M	N	☰	O	⊓	٩

The last nine letters stood for the numbers from 100 to 900.

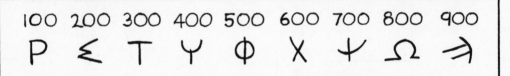

The number 413 in the Ionic
number system looked like this: YAΓ

The Greeks were still not satisfied with their number system.
The third system of numbers they used was called the Herodianic system. It consisted of the following symbols:

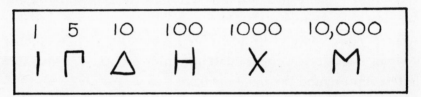

The symbols could be joined together in a certain way to form
the other numbers. The number 5,000 was written this way: ΓX

Fifty looked like this: ΓΔ   Here are some numbers

using the Herodianic numerals:

55	ΓΔ Γ
322	H H H Δ Δ I I
1,632	X ΓΗ H Δ Δ Δ I I

ACTIVITY

Write the following numbers using Herodianic numerals:

67	51
425	6
1,643	2,304

(answers on page 63)

ROMAN NUMBER SYSTEM

The Roman numbers lasted for many centuries. There is still evidence of the use of Roman numerals today — on clock dials, on building cornerstones, in books. The Roman number system consisted of seven main symbols.

The Roman system had one feature that was not found in other number systems at that time. They used the idea of subtraction. If a Roman wanted to write the number four, he placed the symbol for 1 in front of the 5 symbol. This meant that one was to be taken away from the five. If the symbol for 1 was placed after the 5 symbol, then the number was added and was equal to six.

Read the following Roman numbers:

LXVII	MCMXIX	MMCIV
CLXII	MCCCXXIII	MCMLXXIV
MCCXXII	MMCMXVI	(answers on page 63)

ACTIVITY

It is easy to write Roman numerals if you follow the procedure in this example problem. The number 1946 can be written like this:

1000 = M	Write the Roman numeral for
900 = CM	each part of the number.
40 = XL	
6 = VI	
——————	
1946 = MCMXLVI	Then, write the Roman numerals in order.

Write the following numbers using Roman numerals:

964	1,976	1,241
1,872	2,306	3,935

(answers on page 63)

There were many other number systems in use in the ancient world. Some of them used a base of two or five. There are number systems of this kind still in use in some of the uncivilized regions of the world. Modern explorers have found tribes in some areas where no number system is used at all. The tribesmen still express their mathematical ideas in much the same way that prehistoric man did.

HINDU-ARABIC NUMERALS

Our number system in use today was devised by the Hindus in India long before the birth of Christ. The Hindus taught the Arabs the numerals when they traded with them. Later, the numbers were introduced in Europe when the Arabs conquered Spain. The Hindu-Arabic number system took many centuries to develop.

When the numbers were first used, there was no symbol for the zero. At that time, it was impossible to do simple subtraction problems involving borrowing. There was no way to do problems in division and multiplication either.

HINDU NUMERALS 1-10

૧ ૨ ૩ ૪ ૫ ૬ ૭ ૮ ૯ ૦

The number symbols used today are not the same as the ones used in the thirteenth century. The thirteenth-century 5 looked very much like the modern 4.

```
┌─────────────────────────────────────────────────────────────┐
│           CHANGES IN HINDU-ARABIC NUMERALS                    │
│                                                               │
│  12TH CENTURY      1 2 3 ℛ 7 6 ʌ 8 9 0                        │
│  13TH CENTURY      1 2 3 ℛ 4 ɕ ∧ 8 9 0                        │
│  14TH CENTURY      1 2 3 ℛ 5 6 ∧ 8 9 0                        │
│                                                               │
│  PRESENT           1 2 3 4 5 6 7 8 9 0                        │
└─────────────────────────────────────────────────────────────┘
```

Do you wonder why these symbols were selected by the people who invented them? There are many ideas about the origin of the symbols. One idea was that the symbols were selected because the number of angles in the symbol was the same as the number of units it represented. But you can see very quickly that this idea could not be true, because there is only one angle in the thirteenth-century symbol for 7 and not seven angles.

THE ZERO

One of the most important inventions in the Hindu-Arabic number system was the zero. No one is certain when the zero was first used. Some studies show that it may have been used as early as 2,000 B.C. The zero made the place value idea more meaningful and more useful. The number 105 consists of one 100, no tens, and five ones. If the zero did not occupy the tens' place, the number would be 15. The number 105 cannot be written without the zero to hold the tens' place.

Make up a number system of your own. For example, use the following symbols for numbers:

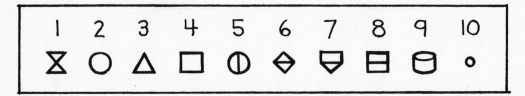

Make up problems with your number system. But first, see if you can do these problems:

$$\begin{array}{c} \mathsf{X}\ \triangle \\ -\ \square\ \diamondsuit \\ \hline \end{array} \qquad \begin{array}{c} \triangle\ \mathbb{O} \\ +\ \boxminus\ \bigcirc \\ \hline \end{array} \qquad \begin{array}{c} \mathbb{O}\ \mathbb{O} \\ +\quad \square \\ \hline \end{array}$$

$$\begin{array}{c} \ominus\ \triangledown \\ -\ \square\ \bigcirc \\ \hline \end{array} \qquad \begin{array}{c} \triangledown\ \diamondsuit \\ -\ \square\ \triangle \\ \hline \end{array}$$

(answers on page 63)

THE ABACUS

The abacus is a device that is used for counting and computing. It consists of a wooden frame with rods or wires. The counters are beads which can be moved along the rod or wire. It was used long before the development of our number system. The Greeks, Romans, Babylonians and Egyptians used various kinds of these counting devices. Perhaps you used an abacus when you were learning how to count.

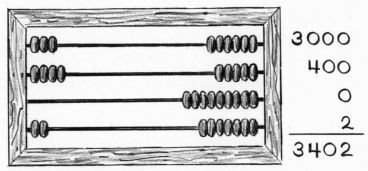

THOUSANDS	3000
HUNDREDS	400
TENS	0
ONES	2
	3402

The amount on the abacus shown equals 3,402. The empty space in the tens' column serves the same function as the written zero.

The word "abacus" has an interesting meaning. It is derived from the Greek word meaning "dust." The first abacus used by the Greeks was a wooden board covered with fine sand. A person would mark the numbers in the sand and rub them out as they were used.

The Roman abacus was constructed in a different way. It has two parts, with two beads in the left-hand section and five beads in the right-hand section. A bead on the left equals five times the value of one bead on the right. Beads are pushed to the center to make the sum, or total.

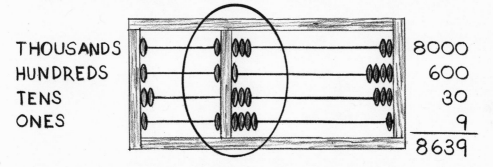

THOUSANDS	8000
HUNDREDS	600
TENS	30
ONES	9
	8639

In the Roman abacus, the left bead in the thousands' column is equal to 5,000. The three beads on the right side in the thousands' column equal 1,000 each, or 3,000. The number on the Roman abacus totals 8,639.

19

You can make a Roman abacus. You will need: 28 beads, four feet of wire, a wooden frame. Make a rectangular frame with two parts as shown in the drawing on the previous page. String the beads on the wire and attach it to the frame. If you do not have beads, you can use washers. Set up the following numbers on your abacus:

27	327	1,250
84	120	1,465

COUNTER RECKONING

During the Middle Ages, the abacus was replaced by the counting board. Counter reckoning was the method of computing in all places of exchange. The tops of tables were ruled with four or five lines, and counters were placed in position to indicate numbers. The word "counter" as used in stores today originated from this business activity of the Middle Ages. The number of counters that had to be used was decreased by having the space between the lines represent 5, 50, 500, and so on.

THOUSANDS	
	500
HUNDREDS	200
	50
TENS	30
	5
ONES	3
	788

The earliest number systems used only whole numbers. Later, fractions were added to express amounts that were less than one whole. The same ten number symbols were used but in a different way. One whole divided into two equal parts was written with one above the two: $\frac{1}{2}$. One whole divided into four equal parts was written as: $\frac{1}{4}$.

The Egyptians were one of the first people to use fractions. However, they used only unit fractions such as $\frac{1}{2}, \frac{1}{3}, \frac{1}{4}, \frac{1}{5}$. The only fraction they knew that did not have a numerator (the part above the line) of 1 was $\frac{2}{3}$. But larger numbers can be written as fractions also. There are 60 minutes in an hour, and 15 minutes is a part or a fraction of an hour. It can be written: $\frac{15}{60}$. For greater simplicity the fraction can be reduced to its smallest numbers. $\frac{15}{60}$ is the same as $\frac{1}{4}$.

Later on, decimal fractions were introduced. They are easier to write or to print on scales and measuring devices. They are used on charts, maps, speedometers and rulers.

Decimals are based on tens, just as the place value of whole numbers is based on tens. Where the number is placed to the right of the decimal point indicates its value. The fraction $\frac{5}{10}$ can be written as a decimal: .5. The 5 is in the tenths place to the right of the decimal point. $\frac{7}{100}$ is the same as .07 because the 7 is in the hundredths place and the zero holds the tenths place. $\frac{354}{1000}$ would be .354.

ACTIVITY

You can find many examples of the uses of decimals and fractions in magazines and newspapers. Have a contest with your friends to see who can find the most illustrations. Give your friends a week to collect examples. You could have another contest to see who can make the longest list of things — such as rulers, thermometers, barometers — that use decimals.

MEASUREMENT THROUGH
THE AGES

During the time of the cave man, there was no need for weights and measures. There was nothing of importance to be measured or weighed. As man learned how to make crude tools and build things, he realized the need for a measure to make his simple construction jobs a little easier. Early man probably used parts of his body to do his measuring. Such a "measuring stick" was always handy.

THE CUBIT

The cubit measure was frequently used by the Egyptians in building their temples and pyramids. It was equal to the distance from the elbow to the finger tips. Since the distance from the elbow to the finger tips of all persons is not the same, the cubit was not a standard measure. Later, the Egyptians did make cubit rods to do their measuring and this standardized the cubit measure to some extent.

THE SPAN

The span was also used by the Egyptians. It was equal to the distance from the thumb to the tip of the little finger of a stretched hand. Two spans are equal to about one cubit.

Take a piece of paper and mark off the distance from your elbow to the tip of your middle finger. This is equal to your cubit. Now measure off two spans on the paper along the line you have marked off. Does your cubit equal two spans?

Take a piece of wood and cut it to the length of your cubit. Measure the length of your bedroom in cubits.

THE INCH

The digit was still another measure used by the Egyptians. It was the smallest one in their table of measures. It was equal to the width of the forefinger. The Romans used a measure that was equal to the width of the thumb.

These two measures led to the inch measure, which was established by Edward II of England in 1324 A.D. He proclaimed that the inch was the distance covered by three barley corns placed end to end.

THE FOOT

The foot was probably the first measure that was used by early man. He had it available whenever he needed it to do his crude measuring. Many thousands of years later, the Greeks also used a foot measure. It was slightly larger than our present-day foot measure.

BARLEY CORNS

← 1 INCH →

The relationship between the inch and the foot was originated by the Romans. They divided the foot into twelve parts. Each twelfth was called a *unicae*. The word "inch" is derived from this word. Today, the foot and the inch are two of the most frequently used measures that we have.

THE YARD

The yard was first used by merchants to measure cloth. It was originally equal to the distance from a man's nose to the tip of his outstretched arm. This measurement caused many arguments because of the different arm lengths of the merchants.

The man with the short arm and big nose got the better deal when he was the seller. His yard was smaller so he got the same amount of money for less cloth. If he was the buyer, he preferred the tall merchant with the longer arm to do the measuring.

In 1439 A.D., the yard was standardized to equal 37 inches. It was not until 1553 A.D. that the legal yard was established as 36 inches long.

THE ROD

In 1514 A.D., the rod was equal to the total length of the left feet of sixteen men, short and tall, who happened to leave the church after services on a certain Sunday. Today the rod is equal to 16½ feet. It is used as a unit of land measure.

THE FATHOM

The fathom was used by Anglo-Saxon sailors in Britain. They would drop a rope with a stone attached to it in the water to determine the depth. As they took the rope out of the water, they would stretch their arms and count the number of arm lengths, or fathoms. The fathom was equal to approximately six feet. Today it is accepted as being equal to exactly six feet.

THE PACE

Those of you who are in the Boy or Girl Scouts are familiar with the Scout pace. Do you have any idea where the pace originated? The pace, as you may know, is equal to a double step. The Roman army officers were the founders of the pace. A Roman soldier's pace was equal to about five feet—from the heel of one foot to the heel of the same foot when it next touched the ground.

Roman legions were famous for their marching ability. The Roman officers kept a record of the number of paces from one military outpost to another. This information made it possible for them to estimate, very accurately, how long it would take to march to a certain destination.

THE MILE

The word "mile" is derived from the Latin word *mille* which means "thousand." The Roman mile was equal to 1,000 paces. *Mille passuum* in Latin means a thousand paces.

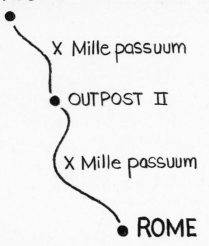

OUTPOST I

X Mille passuum

OUTPOST II

X Mille passuum

ROME

Here is a Roman officer's log book. The number of paces from Outpost I to Outpost II is equal to 10,000. The number of paces from Outpost II to Rome is equal to 15,000.

Answer these questions:

1. If the Roman soldiers could cover 4,000 paces in an hour, what time would they arrive at Outpost II if they started at 9 A.M. from Outpost I?
2. How long would it take the soldiers to march from Outpost II to Rome if they marched 3,000 paces an hour?
3. What time would they arrive in Rome?

(answers on page 64)

←PACE→

Measure your pace by counting how many double steps it takes you to cover 20 feet. Divide the number of double steps into 20 to find the length of your pace.

Pretend that you are a Roman officer. Count the number of paces from your home to school. Using the second hand of a wrist watch, determine how long it takes you to walk ten paces. How long should it take you to walk to school?

THE METRIC SYSTEM

The metric system is widely used in many countries today. The meter was first proposed to the French government in 1799, but it was not until 1840 that it became obligatory to use the metric system in France. Although the United States uses the English system of measure, American scientists use the metric system because it is easier to make precise computations with it. The meter is equal to 39.37 inches. It can be divided into hundredths and thousandths. Each centimeter (.3937 inches, or one hundredth part of a meter) on the scale is divided into millimeters. A millimeter is equal to .03937, or one thousandth part of a meter.

The kilometer is used to measure great distances. Road signs in Europe are marked in kilometers instead of miles. A kilometer equals about 0.6 of a mile, or exactly 1,000 meters.

Metric System

1 meter equals 39.37 inches
10 millimeters   =   1 centimeter
10 centimeters   =   1 decimeter
10 decimeters   =   1 meter
1,000 meters   =   1 kilometer

ACTIVITY

Make a meter stick by cutting a piece of wood about 39½ inches long. (39.37 equals approximately 39½ inches.) Measure the length of your room with your meter stick. About how many decimeters long is your room? (Remember that ten decimeters equal one meter.) Measure the width of your room. About how many decimeters is the width?

27

LARGE AND SMALL
MEASUREMENT

Today, measurements can be made to the millionth of an inch. Distances from the earth to the stars can be calculated in the trillions of miles. The scientists of long ago did not have the equipment to measure such large and small distances. Galileo (1564-1642 A.D.) was a famous Italian scientist who tried to measure the speed of light but was unsuccessful because he lacked proper instruments.

A LIGHT YEAR

The sun is the closest star to the earth. It is about 93 million miles away. Alpha Centauri is the next nearest star. It is about 26 million million (26 trillion) miles away. Scientists do not use miles as the measure for such great distances. Instead, the light year is used. Light travels very fast. It moves at the speed of about 186,000 miles in a second.

A light year is the distance light travels at that speed in a year. It is equal to six million million miles.

Alpha Centauri is 4.3 light years away. When you look at this star, you see it not as it is today, but as it was 4.3 years ago. It takes that long for the light which enables you to see it to travel from this star to your eyes. There are many stars in the universe that are thousands and even millions of light years away.

EARTH   4.3 LIGHT YEARS FROM ALPHA CENTAURI

Some stars are brighter than others. Astronomers use numbers to compare the brightness of one star with another. The numbers are called magnitudes. A star which is classified as having zero magnitude is 2½ times brighter than a star of the first magnitude. A star of the first magnitude is 2½ times brighter than a star of the second magnitude, and so on. The brighter the star, the less its magnitude number. Stars up to the sixth magnitude can be seen with the naked eye. By using a telescope, stars of the 20th magnitude can be seen.

THE MICROMETER

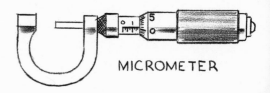

MICROMETER

The micrometer is an instrument used to make very small measurements. You can measure the thickness of an object to the nearest thousandth of an inch with a micrometer, which consists of two scales, one on the barrel and another on the thimble. Each mark on the barrel equals 0.025 (twenty-five thousandths) of an inch. Each mark around the thimble equals 0.001 (one thousandth) of an inch.

The reading in the illustration equals 0.345, or 345 thousandths of an inch. Micrometers are used to measure airplane and missile parts, among other things.

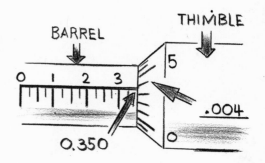

0.350   BARREL READING
+.004   THIMBLE READING
─────────
0.354

0.025

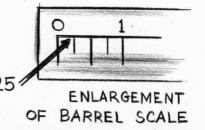

ENLARGEMENT
OF BARREL SCALE

EXPONENTS

Large numbers can be written in a short way by using exponents. A number like $4 \times 4 \times 4$ is written 4^3. The small 3 is called an exponent. It tells the number of times you multiply 4 by itself. $4 \times 4 \times 4 = 64$ and $4^3 = 64$.

The number 100 can be written as 10×10 or as 10^2. Exponents can be used to write any number. The number 5,000 can be written as $5(10)^3$ which is the same as $5 \times 10 \times 10 \times 10$ which equals 5,000. The parentheses mean to multiply, also.

A number like 1,736 can also be written using exponents if it is broken down into separate parts called "terms."

$$
\begin{array}{rcl}
1000 & = & 10^3 \\
700 & = & 7(10)^2 \\
30 & = & 3(10) \\
6 & = & 6 \\
\hline
1736 & = & 10^3 + 7(10)^2 + 3(10) + 6
\end{array}
$$

The number 1,736 consists of four terms when written with exponents.

ACTIVITY

1. Borrow a micrometer from an engineer. Measure the thickness of the following objects: a pencil, five sheets of paper, a paper clip, a rubber band, an eraser, a comb.
2. Write the following numbers using exponents.

462	1,340	3,156
511	3,160	7,540

(answers on page 64)

The early Egyptians built pyramids and temples that have stood for many centuries. Some of the stone blocks that were used in building the pyramids weighed as much as thirty tons. Since there were no machines to lift the blocks, it took many men to move the huge stones into position.

The Egyptians managed to do it, and very accurately, too. They had a working knowledge of practical geometry. They used the right angle—the corners of your book form right angles—to square the corners of the pyramids and temples.

Angles are measured in degrees. The minute hand of a clock travels 360 degrees in one hour. It moves 180 degrees in a half hour. A right angle contains 90 degrees. It is one quarter of a complete circle, and any circle has 360 degrees. The symbol for a degree is a tiny circle. Ninety degrees is written: 90°

ROPE STRETCHERS

A rope knotted into twelve equal sections was used to lay off a right angle. One rope stretcher, called a *harpedonapta*, held the knots at each end of the rope. He then placed the rope at the point (A) where the right angle was to be formed.

Another rope stretcher held the knot that was four equal sections away from A. The third man held the knot that was three equal sections away from A. By stretching the rope tightly, a right angle was formed.

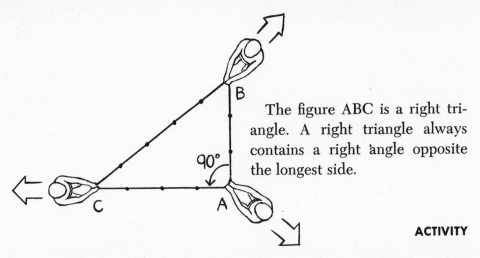

The figure ABC is a right triangle. A right triangle always contains a right angle opposite the longest side.

Measure off twelve equal sections on a rope. Mark the end of each equal section with a knot or adhesive tape. Have two friends help you as rope stretchers. Lay off a right angle like the rope stretchers of ancient Egypt did.

AREA OF A TRIANGLE

The Egyptians knew how to find the area of a right triangle. The formula for the area of a triangle was found in the Ahmes Papyrus. It stated that the area of any triangle was equal to one-half of the base multiplied by the side, or $\frac{1}{2}(b \times s)$.

Here is how the Egyptians derived the formula.

Figure ABCD (fig. 1) is a rectangle. A rectangle has opposite sides equal and parallel. It contains four right angles of $90°$. The area of a rectangle (or of a square, too) equals the length or base multiplied by its side or height (b × h). If b = 6 feet and h = 4 feet, then the area of the rectangle ABCD is 6×4 or 24 square feet.

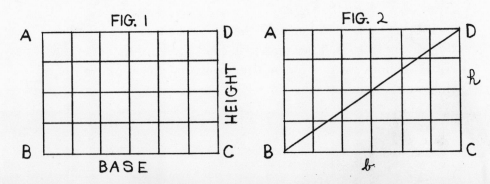

By drawing line BD, two equal triangles are formed (fig. 2). Each one equals one-half the rectangle, and so the area of each triangle equals one-half the area of the rectangle or $\frac{1}{2}$ (b × h).

The two triangles are right triangles, and it so happens that the side of a right triangle is the same as its height. But this is not true of other kinds of triangles. The Egyptians did not realize that their formula applied only to right triangles if they used the side instead of the height. $\frac{1}{2}$ (b × s) is not always the same as $\frac{1}{2}$ (b × h).

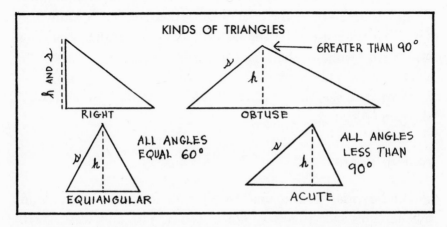

The area formula for *any* triangle can be derived from a parallelogram. In the figure below, ABCD is a parallelogram (fig. 1). It does not have right angles, but the height (h) makes a right angle of 90° with the base. By drawing line DB, the parallelogram is divided into two equal triangles (fig. 2).

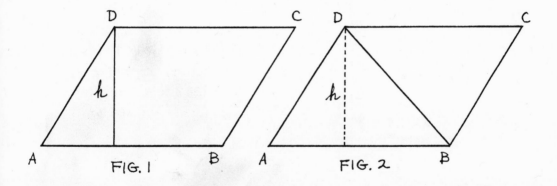

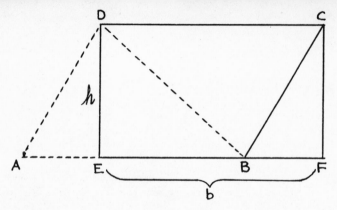

The parallelogram can be changed into a rectangle by moving the right triangle on the left to the right of the figure.

This does not change the total area of the parallelogram, and its area is the same as the area of the rectangle that is formed, or (b × h). Since line DB originally divided the parallelogram into two equal triangles, the area for either triangle will be half the total area, or $\frac{1}{2}$(b × h).

As you can see, the height and side of a triangle is not always the same distance.

FINDING NORTH

The Egyptian gods demanded that the four sides of pyramids and temples face precisely north, south, east and west. North was located by observing the path of the sun from sunrise to sunset as it moved from east to west. A stake was placed in the ground. As the sun rose, the end of the stake's shadow was marked (A) and as it set, another mark (B) was made.

By swinging an arc of the same radius with a piece of rope from point A and from point B, point C was found, the place where the two arcs met. A straight line from the stake to point C located north.

RADIUS OF CIRCLE

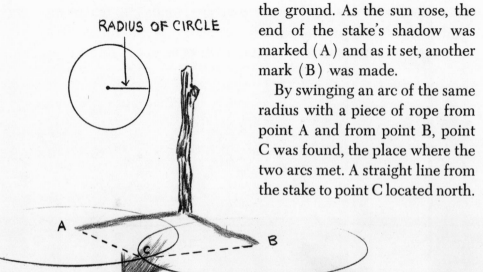

LOCATING BOUNDARIES

It was fortunate for the Egyptian landowners that the pyramids and temples were situated in a precise direction. The Nile River overflowed its banks each year. Landmarks and boundary lines were washed away. The temples and pyramids served as permanent reference points for relocating landmarks and boundaries. This is one way that boundary lines were relocated after the floods.

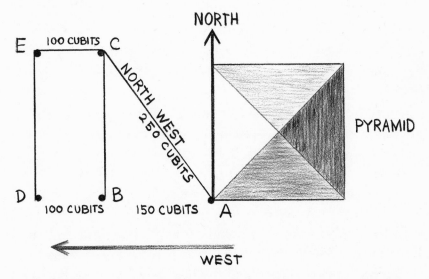

By knowing the distance from the corner of the pyramid A and determining the direction of northwest, landmark C was located. Landmark B, which was directly west of the corner of the pyramid, was located in the same way. Landmarks D and E were found by using the other landmarks as reference points and knowing the distance between them before the boundaries were washed away.

Not all fields or boundaries were rectangular in shape, but odd-shaped ones could be measured off as triangles. With their knowledge of triangles the Egyptians could survey any land.

PYTHAGOREAN THEOREM

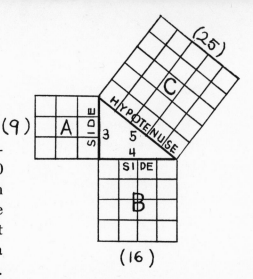

Pythagoras was a Greek mathematician who lived over 2,500 years ago. He was very much interested in the right triangle and he made a very important discovery about it. He made a drawing like the one at the right.

No matter what length he made the legs of the right triangle, the number of squares in A plus the number of squares in B equaled the number of squares in the hypotenuse C. (The hypotenuse is the longest side of a triangle.) Another way of writing this was a formula. (Remember that the small 2 means to multiply a number by itself.)

$$\text{side}^2 + \text{side}^2 = \text{hypotenuse}^2$$
$$\text{or} \quad 3^2 + 4^2 = 5^2$$
$$\text{or} \quad 9 + 16 = 25$$

This formula works for any right triangle. The rope stretchers of Egypt who lived before Pythagoras may have understood this special fact about right triangles when they laid out their right angles, but it was the Greeks who proved it and also many other facts of geometry that we still use today.

ACTIVITY

Draw a right triangle with a 6-inch side and an 8-inch side. Without measuring the hypotenuse, can you find the length of the hypotenuse? Remember that $6^2 + 8^2 = \text{hypotenuse}^2$. Try experimenting with other right triangles.

(answer on page 64)

Before any complicated object is constructed, a scale drawing of the object is often made. A scale drawing is usually smaller in size than the object itself, but it is the same shape and proportion. Such scale drawings are usually drawn on graph paper. This paper is divided into small squares which make it easier to make exact measurements.

Scale drawings are also used by engineers to find heights and distances that cannot be measured directly. It would be very difficult to measure the base of a mountain with a tape measure, but it can be done indirectly.

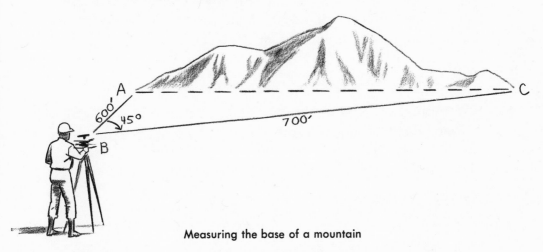

Measuring the base of a mountain

The surveyor in the illustration is finding the distance from A to C indirectly. By means of a transit, an instrument used to find angles, he knows that angle B is 45 degrees. The distance BA is equal to 600 feet, and the distance BC equals 700 feet. These distances can be measured. At his desk, the surveyor uses this information to make a scale drawing on graph paper.

Each square on the graph paper is ¼ inch long. He decides that each square will represent 50 feet. Therefore, four squares will make one inch, and one inch will equal 200 feet. A line 3 inches long is drawn to represent BA which is 600 feet. A line

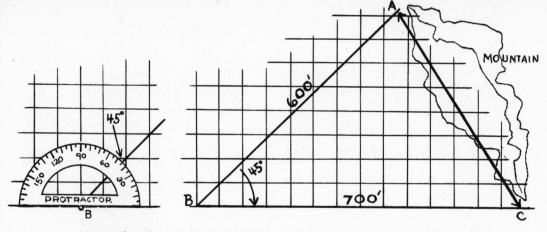

3½ inches long is drawn to represent BC which is 700 feet. The angle between is 45 degrees. Now, by measuring the length of line AC, the approximate length of the base of the mountain is found. It is equal to 2½ inches, or 500 feet.

ACTIVITY

You can make a hand transit to find angles. Cut a circle with a 7-inch diameter out of cardboard. (The diameter is the distance through the center of the circle.) Divide the circle into 360 degrees (fig. 1) for your transit scale.

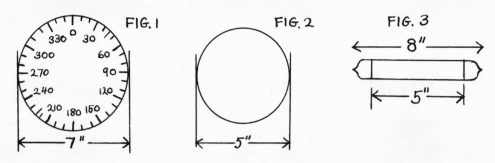

Cut another circle with a 5-inch diameter. Cut a strip of cardboard one inch wide and 8 inches long (fig. 2). Make the ends pointed as shown, and fold up each end so that 5 inches remain in the center. Staple the strip to the 5-inch circle. Place the larger circle under the smaller one and secure them together with a paper fastener (fig. 3).

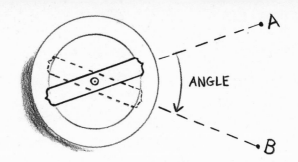

ANGLE

Now you are ready to use your hand transit. To find the degrees of an angle between points A and B, line up the two paper points with the zero on your scale and point A. Without moving the outer circle, swing the part with the pointed ends around and sight on point B. The number of degrees on the scale will be the angle. A surveyor's instrument is more precise, but you can measure angles fairly accurately with your hand transit.

You can use your transit to lay out a baseball field. You will need: four stakes, your hand transit, a friend.

A baseball diamond is a square with each side equal to 90 feet. Each angle of the square equals 90 degrees. Insert a stake in the ground. This will be home plate. Measure off 90 feet. Stick a stake in the ground at this point, which will be first base. Stand on first base and sight home plate. Now swing the sight of your transit 90 degrees, and have your friend measure 90 feet along the 90-degree angle. Stick another stake in the ground. This is second base.

Stand on second base, sight first base, and find the 90-degree angle for third base. Have your friend measure 90 feet and mark third base. The remaining distance from third to home plate should also be 90 feet. If you have come within five feet, consider yourself an expert junior surveyor.

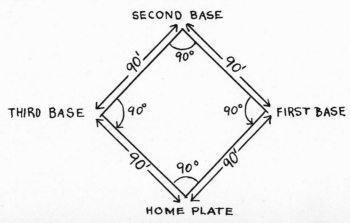

SECOND BASE

90°

90'   90'

THIRD BASE   90°   90°   FIRST BASE

90'   90'

90°

HOME PLATE

ESTIMATING WITH TRIANGLES

RATIO

Groups of objects can be compared in different ways. If there were 20 boys and 10 girls in a class, you could compare them in the following ways:

There are 10 more boys than there are girls.

There are 10 less girls than boys.

There are half as many girls as there are boys.

In the last statement, the numbers were compared by dividing the number of boys into the number of girls ($10 \div 20 = \frac{10}{20} = \frac{1}{2}$). When two numbers are compared by division, it is called *ratio*.

Lines can be compared by ratio. A line three inches long can be compared to a line six inches long by the ratio of 3 to 6, or $\frac{3}{6}$. By reducing the fraction, the ratio becomes $\frac{1}{2}$. A line three inches long is $\frac{1}{2}$ as long as a line six inches long.

3"

6"

SIMILAR TRIANGLES

Have you ever looked at the drawing of a triangle through a magnifying glass? If not, borrow one and try it. Draw a triangle on a piece of paper. Look at it through the magnifying glass.

You will see that the image of the triangle is much larger than the drawing. But the magnifying glass does not change the *shape* of the triangle. Triangles that are the same shape are called similar triangles. When two triangles are similar, the parts in the same relative position are called corresponding parts.

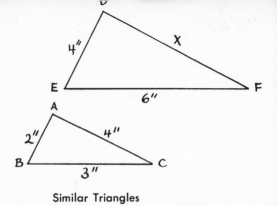

Triangles ABC and DEF are similar. Side AB is two inches, and its corresponding side DE is four inches. AB is half as long as DE, and the ratio of AB to DE is 1 to 2, or $\frac{1}{2}$.

Similar Triangles

This ratio of the two shortest sides of the two similar triangles can be written as $\frac{AB}{DE}$ or $\frac{2}{4}$ or $\frac{1}{2}$

PROPORTION

A statement that two ratios are equal is called a *proportion*. In a proportion such as $\frac{3}{6} = \frac{2}{4}$ the "cross products" are equal. This means that the amounts diagonally across, when multiplied, are equal.

$$\frac{3}{6} \diagup\!\!\!\!\!\diagdown \frac{2}{4} \qquad\qquad 3 \times 4 = 12$$
$$2 \times 6 = 12$$

You can use this rule to find the unknown sides of triangles.

In triangle DEF above, the length of side DF is not given. But since any pair of corresponding sides of similar triangles are in proportion, then DF can be found by setting up the ratios of corresponding sides and a proportion: $\frac{AB}{DE} = \frac{AC}{DF}$

We know that AB = 2, DE = 4, and AC = 4. By substituting these numbers and multiplying, we get the cross products which we know will be equal.

$$\frac{2}{4} = \frac{4}{DF} \qquad\qquad \frac{2}{4} \diagup\!\!\!\!\!\diagdown \frac{4}{DF} \qquad\qquad 2 \times DF = 4 \times 4$$
$$2\,DF = 16$$

If 2 DF equal 16, then one DF equals 16 divided by 2, or 8. You can check this answer easily. Side AC should be half as long as side DF and 4 is half of 8.

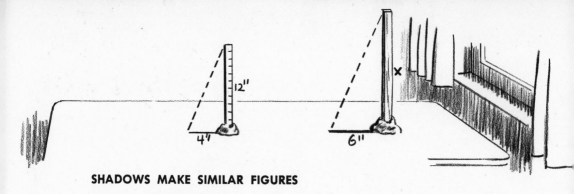

SHADOWS MAKE SIMILAR FIGURES

With your knowledge of similar triangles, you can use shadows to estimate the height of a building, a tree, a stick. Try this experiment. Place a 12-inch ruler and a stick of unknown length upright on a table near a window. The sun must be shining so that shadows will be made. Use some clay to make the ruler and stick stand upright.

If you could imagine a line connecting the tops of the stick and the ruler to the ends of their shadows, similar right triangles would be formed. By measuring the length of the shadows a ratio can be set up. The ratio of the two shadows in the illustration is 4 to 6. The ratio of the ruler and the stick is 12 inches to X. Since the corresponding sides of similar triangles are in proportion, you can find the length of X by finding the equal cross products.

$$\frac{12}{X} = \frac{4}{6} \qquad\qquad \begin{aligned} 4 \times X &= 12 \times 6 \\ 4X &= 72 \\ X &= 18 \end{aligned}$$

ACTIVITY

Use the same method as above to estimate the height of a small tree. Your problem would look like this:

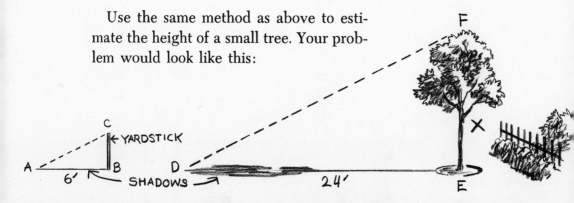

In the similar triangles ABC and DEF, the corresponding sides are in proportion. You can measure the length of the shadows. If AB = 6 feet and DE = 24 feet, then you have:

$$\frac{AB}{DE} = \frac{BC}{EF} \text{ or } \frac{6}{24} = \frac{3}{EF}$$

$$6 \times EF = 24 \times 3$$
$$6EF = 72$$
$$EF = 12$$

The height of the tree (EF) is 12 feet.

ESTIMATING HEIGHTS WITH A MIRROR

You can also estimate heights with a mirror. Place a mirror on the floor. Move back until you see the reflection of the ceiling corner in the mirror. Two similar triangles are formed. The corner of the room forms one side of one triangle. Your height forms a corresponding side of a similar triangle.

If you are five feet tall, then you know the distance DE. Distances BC and DC can be measured. If BC equals 6 feet and DC equals 3 feet, then you have the ratio and proportion:

$$\frac{X}{5} = \frac{6}{3}$$

$$3X = 5 \times 6$$
$$3X = 30$$
$$X = 10$$

In a proportion, the cross products are always equal. Thus, the height of the room is 10 feet.

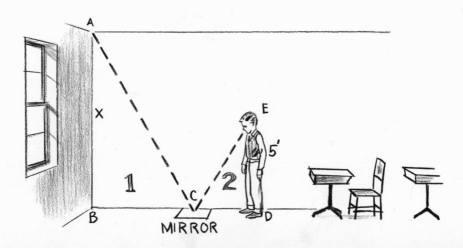

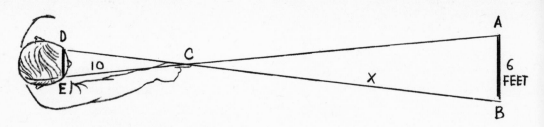

ESTIMATING GREAT DISTANCES

It has been found that the distance from a person's eye to the end of the middle finger of his outstretched hand is approximately ten times the distance between the eyes.

By closing the right and then the left eye, the end of the outstretched finger appears to move from point A to point B. Suppose AB is estimated to be equal to 6 feet.

Triangles DEC and CAB above are similar. In the proportion $\frac{DE}{AB} = \frac{EC}{CB}$ you can substitute the known facts:

$$\frac{1}{6} = \frac{10}{X} \qquad\qquad \begin{array}{r} 1 \times X = 10 \times 6 \\ X = 60 \end{array}$$

The distance from C to B is 60 feet.

The next time you want to estimate how far a ship is from shore, you can use similar triangles. Sight the ship by closing the right eye and pointing to the ship. Now close the left eye and observe that the ship appears to move from A to B. Estimate the distance AB by judging the number of ship lengths it is equal to. (You will have to guess at the ship's length and the number of ship lengths from A to B.) If the distance AB is estimated to be 200 feet, the ship is ten times that, or 2,000 feet from shore.

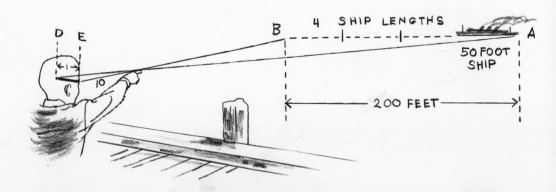

Long ago there was no need to know the exact time of day. There were no trains to catch, no appointments to keep, and no tardy time for school. The hours of the day were not known to early man. When the sun went down in the sky, he knew it was time to build a fire for the night to keep wild animals away. When the sun rose in the morning, his stomach told him it was time to eat.

Later on, man began to realize that time was important to him. Once he had invented number systems, he could keep an accurate record of his time.

No one really knows when man first started to keep time with some sort of an instrument. Maybe he first observed how a shadow formed by a large rock changed its position during the day. Perhaps he placed small rocks around a large rock to divide the daytime into parts. Each rock could have represented something he had to do during the day.

GNOMONS

The earliest sundial was made in about 1,500 B.C. by the Egyptians. Called a gnomon, it was a rod or stone shaft on a level surface. The dial face was the ground. The shadow cast by the gnomon would move as the sun trav-eled from east to west. The Babylonians, who were very advanced in mathematics and the study of the heavens, probably used some form of a sundial, too.

GNOMON

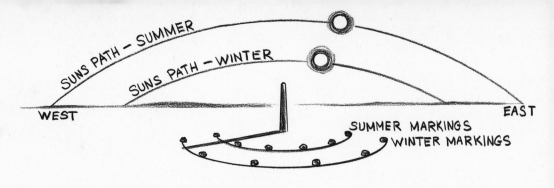

SUNS PATH — SUMMER

SUNS PATH — WINTER

WEST

EAST

SUMMER MARKINGS

WINTER MARKINGS

THE SUNDIAL AS A CALENDAR

The Egyptians observed that the shadow of the gnomon pointed in a different direction during the day. They marked the ground for their hours. Then they noticed that the shadow was a different length in Spring than it was in Winter.

They also observed that the sun did not rise or set in the same locations during the year. The shadow was shorter in the summer than it was in the Winter. This led to the development of a number of different marks for their hours and seasons. Thus, the Egyptian gnomon was used as a calendar as well as a clock.

OTHER SUNDIALS

The Egyptians also developed a portable sundial. The upper part of the T-shape cast a shadow on the strip of wood that was marked off in segments. As the sun moved, the shadow marked the hours. In the morning the crossbar faced east. After the noon hour, the sundial was turned so that the crossbar faced west.

Circular sundials were used by many early civilizations, and they are still in use today. A stationary pointer, or gnomon, points to the North Pole in the northern hemisphere or to the South Pole if the sundial is south of the equator. This kind of sundial was a favorite time piece of the early Romans, and Roman numerals are still used on the faces of circular sundials.

SUN'S RAYS

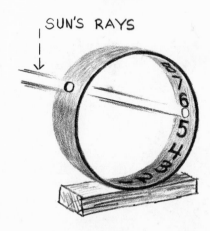

SUN'S RAYS

Berossus, a Chaldean astronomer who lived about 300 B.C., used a hollow half-sphere to record the time of day. The shadow of a bead at the center of the sphere moved around the inside as the sun moved from east to west. At that time, the day was divided into twelve equal parts.

There were other circular sundials. A ring-shaped dial told time by the sun's rays directed through a hole. The daylight hours were marked inside the ring.

ACTIVITY

You can make a T-shaped sundial like the Egyptians used 3,000 years ago. You will need a piece of wood 15 × 1 × ½ inches. Use a ruler to mark the stick at points four inches and six inches from one end. Saw the stick at these two points. Now nail the three pieces together as shown in fig. 1.

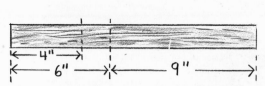

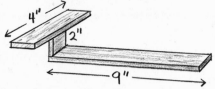

Place your sundial in the sun so that the crossbar faces east if it is morning, or west if it is afternoon. Check the time with a watch. On the hour, draw a line where the shadow of the cross-bar falls on the 9-inch piece which will be your ruler-type dial face. You can mark the other daylight hours by checking the shadow on the dial at hour intervals.

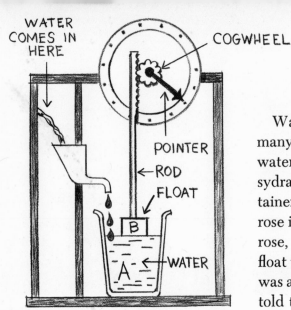

WATER COMES IN HERE

COGWHEEL

POINTER

ROD

FLOAT

B

WATER

A

WATER CLOCKS

Water clocks were used by many ancient peoples. The Greek water clock was called the "clepsydra." Water dripped into a container (A). As the water slowly rose in the container, the float (B) rose, too. The rod attached to the float turned the cog wheel, which was attached to the pointer which told the time.

THE SANDGLASS

The sandglass has been in use for many centuries as a means of measuring time. Perhaps you have used one to boil "three minute" eggs for breakfast. You know that it takes three minutes for the sand to trickle through the narrow opening into the bottom part of the glass. The sandglass, or hour glass, of long ago was made so that it took one hour for the dry sand to drop into the lower part of the glass.

CANDLE AND ROPE CLOCKS

A candle clock was a popular timekeeper before mechanical clocks were invented, and rope clocks were also used at one time. By marking notches on a candle, or making knots in a piece of rope, people could tell time by noting how many sections had burned away.

COAT HANGER

OPEN END

HOLE PUNCHED IN BOTTOM

JUICE CAN

GLASS

You can make a sandglass by using things that you can find at home. You will need: a small juice can, a coat hanger, a glass, some sand or salt. Punch a hole in the bottom of the juice can. Shape the coat hanger into a stand for the can.

Place the juice can on the stand. Now set the glass under the can. Fill the can with sand or salt. Record the length of time it takes for the sand or salt to drop into the glass. If it takes one-half hour, then you have made a one-half hour sandglass. The size of the hole will determine how slowly or how fast the sand or salt will run out.

MODERN CLOCKS

Later on, mechanical clocks were invented. The pendulum clock—like a grandfather's clock or a cuckoo clock—was devised, and today we have watches and clocks run by a spring and balance wheel. But even more accurate time pieces have been invented. The atomic clock which uses the motion of ammonium molecules to make it run is the most accurate in existence.

As man discovered number systems, they were soon put to use on his time pieces. The crude sundials of long ago have now been replaced with high-precision timekeepers. Numbers have aided man to keep a more accurate account of his time.

Today scientists calculate lengths of time in thousandths of a second. Astronomers can determine within a fraction of a second when a certain star will appear at a predetermined point in the sky. None of this would be possible without a knowledge of mathematics.

MATHEMATICS BUILDS
MANY THINGS

The city can be a very confusing place with its fast-moving cars, tall buildings, and its symphony of off-beat sounds. Yet in this busy surrounding there exists a world of mathematics which has a constant influence on our lives. Mathematics has been called the "queen of the sciences," which is not surprising when so many things depend on it.

Automobiles and the many other kinds of vehicles would not be in existence today if it were not for mathematics. The tall skyscrapers were made possible because engineers used mathematics to draw the plans, make scale drawings, and determine the kinds and amounts of materials needed.

Builders of long ago had only very crude tools, but they had plenty of manpower and a knowledge of mathematics. The Egyptian engineers who directed the building of the pyramids sometimes used as many as 30,000 men to raise the huge blocks into place. Ancient structures hundreds of feet high have lasted because the mathematical calculations were accurate.

The Greeks constructed beautiful buildings whose styles are still imitated today. They relied on mathematics to make their buildings precise in every detail. The Roman engineers constructed the famous Appian Way for the Roman Legions, as well as other roads, bridges and aqueducts. The Great Wall of China, whose length would stretch halfway across the United States, was built many centuries ago. Without a knowledge of mathematics, none of these building feats would have been possible.

BRIDGES

All engineers use mathematics in many ways. The men who plan and build bridges must know a great deal about mathematics. It takes many hundreds of pages of mathematical calculations to build a bridge, and all must be exactly right. If the foundation is not properly poured or the cables are weak, the bridge would soon collapse. The engineer relies on the triangle, a geometric figure, to make the bridge sturdy. A triangle is very rigid. Much of the steel work for bridges is riveted together in triangular shapes.

ACTIVITY

You can test the rigidity of a triangle. Cut four strips of cardboard eight inches long and two inches wide. Punch a hole in each end. Make a rectangle by fastening the four strips together with paper fasteners as shown in the drawing. Pull on the rectangle and observe how it moves freely.

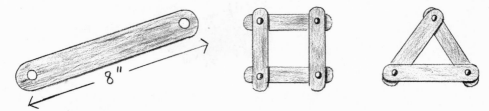

Now take the rectangle apart and form a triangle with three of the pieces. Pull on the triangle. Observe how rigid it is. This is the reason for its use in building bridges.

SKYSCRAPERS

The first skyscraper was a tower called a *ziggurat*. Built by the Babylonians thousands of years ago, ziggurats were shaped like a pyramid and built in many levels. At the top was the shrine of a god where priests conducted worship. This ancient skyscraper, made of brick, was about the height of a modern five-story building. The men who built it knew enough of mathematics to determine the strength of the building materials used, and they also had a working knowledge of right angles.

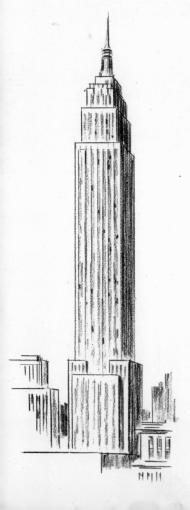

Today, our modern skyscrapers go up many times higher than the Babylonian ziggurat. Many months of careful planning are necessary before actual construction begins. The foundation must be a certain depth so that it will support the building. The amount of cement or steel, the number of girders and their exact size, and thousands of other details have to be determined accurately. Mathematics provides the answers for all these problems.

The Empire State Building in New York City stands 1,250 feet above the ground. Rising 102 stories, it was completed in less than a year. All the steel pieces that were used in the building were made and marked in the factory, then fitted together like a giant jig-saw puzzle. Scale drawings were used by the thousands. Each step involved mathematics.

You can build a skyscraper out of tooth-picks. You will need: a box of toothpicks, airplane glue, clay. Form the clay into a square about six inches on each side. Use this as your foundation. Stick nine toothpicks into the clay as shown in the drawing. Now start erecting your skyscraper by gluing the ends of the toothpicks together. See how high you can make your building.

SATELLITES AND MISSILES

Today there are many artificial satellites circling the earth. But before those satellites were put into orbit, countless mathematical calculations were needed. The entire science of astronautics depends very much on mathematics.

Timing, speed or velocity, direction, altitude, amount of fuel for proper thrust, orbital paths—all details must be plotted and planned well in advance of any space flight. The missile used for launching must function accurately at every stage. An astronaut's capsule must be in perfect working order. Measurements have to be made to the minutest degrees and the smallest of fractions.

Before America's first astronaut, Lt. Col. John Glenn, rocketed into orbit in the "Friendship 7" an astounding number of mathematical jobs had to be done. Yet the "Friendship 7" will soon be a relic in the annals of space flight as, with the help of numbers, man goes on to land on the moon and visit other planets. In the conquest of outer space, the value of mathematics cannot be overrated.

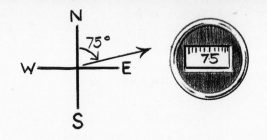

THE COURSE

An airplane pilot is able to fly on invisible roads in the sky because of his knowledge of mathematics. The compass and the protractor are two of the mathematical instruments used by a pilot or navigator. The protractor is used to determine the course. The compass, located on the instrument panel, tells the pilot if he is on that course. The compass reading above is 75 degrees. Compass degrees indicate direction clockwise from north. Seventy-five degrees is a north-easterly direction.

AERONAUTICAL CHARTS

Before a pilot takes off, he determines his compass course by plotting his flight on an aeronautical chart. These charts can be obtained for any area. They include a circle divided into 360 degrees with a rose to indicate true north direction.

If a pilot wishes to fly from Bakerville to Seaview, he would first draw a line from Bakerville to Seaview. A line would also be drawn directly north from Bakerville. Then he would find the number of degees in the angle that is formed by using the protractor as shown in the drawing. If this angle is equal to 75 degrees, he will hold his plane's compass to 75 degrees.

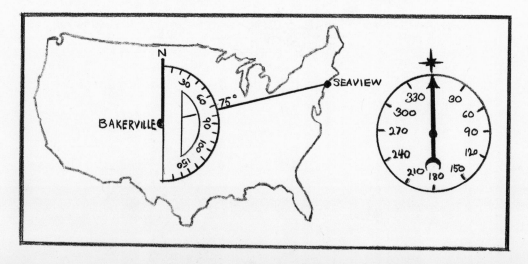

CORRECTING THE COURSE

The compass in an airplane, like any compass, works because the earth is a huge magnet, with two magnetic poles, one located in northern Canada and the other in the Antarctic Ocean. A compass indicates magnetic north, which is not always the same as true north as shown on a map.

Because of this, the pilot or navigator must check his charts and maps and make a compass correction to determine the real course to fly. By using mathematics, this is easily done.

AIR SPEED

After the course has been corrected, the wind direction has to be checked. Wind can easily blow an airplane off course.

Maybe you have heard a pilot say, "We rode a tail wind." He meant that the wind was traveling in the same direction as the plane. In the illustration, a jet airliner is traveling 600 miles an hour. This is the air speed of the jet plane. But a tail wind is blowing 20 miles an hour. The speed of the airplane over the ground is 620 miles an hour.

If a 20-mile-an-hour wind was blowing at the nose of the jet plane, the ground speed would be only 580 miles an hour.

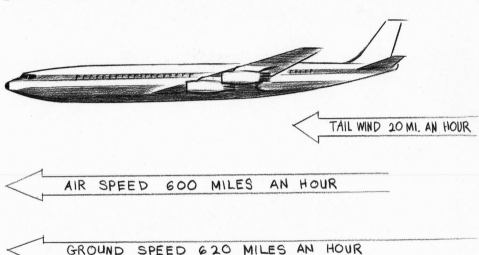

TAIL WIND 20 MI. AN HOUR

AIR SPEED 600 MILES AN HOUR

GROUND SPEED 620 MILES AN HOUR

BLOWING OFF COURSE

The wind does not always blow at the nose or the tail of an airplane. Very often it blows the airplane to the right or to the left of the desired course.

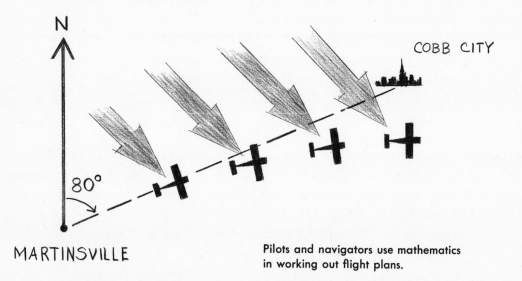

Pilots and navigators use mathematics in working out flight plans.

In the illustration above, a light cabin plane sets out from Martinsville to Cobb City, a distance of 240 miles, flying a course of 80 degrees. The constant force of the wind blowing in a south-easterly direction blows the airplane off course.

But before the pilot began his flight, he worked out a mathematical problem to determine the direction he must fly to compensate for the wind. By allowing for wind direction and speed, he can adjust his compass course so that he will arrive at Cobb City and not south of it.

Mathematics plays a very important role in flying. Not only is the plane's course figured out mathematically, but fuel consumption, time schedules, design of airplanes and many more jobs in aviation involve the use of mathematics.

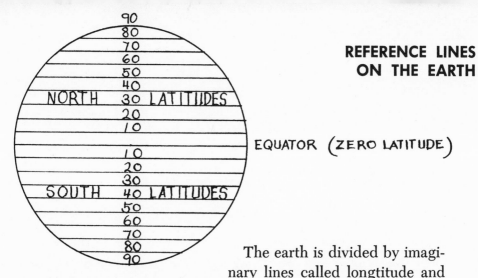

The earth is divided by imaginary lines called longtitude and latitude. The lines of latitude run parallel to the equator. Those of longitude run from the North Pole to the South Pole. Longitude and latitude are measured in degrees. A circle has 360 degrees, and the almost-round earth is divided into 360 degrees.

LATITUDE

Latitude lines that are north of the equator are north latitude. Those south of the equator are south latitude. A person standing at the equator would be at zero degrees latitude. The poles of the earth are 90 degrees latitude, since the distance from the equator to either the North Pole or the South Pole is one-fourth of a complete circle around the earth.

Many maps show latitude lines, but the navigator of a ship uses an instrument called a sextant to find the ship's latitude. The sextant measures the angle formed by the horizon and Polaris, the North Star. This angle is the latitude position of the

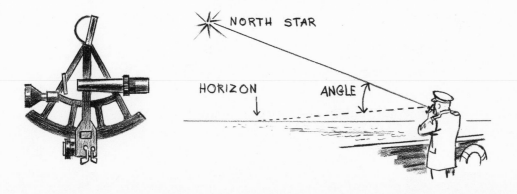

ship. It can also be found by sighting on the sun and the horizon at noon. If the navigator was at the North Pole, Polaris would be directly over his head and the sextant would read 90 degrees. At the equator, Polaris is on the horizon, and the reading would be zero degrees.

ACTIVITY

You can find your latitude by using the two stars in the scoop of the Big Dipper as a measuring rod. The distance between the two stars is equal to five degrees. By estimating how many of these lengths it is from the North Star to the horizon, you can obtain your latitude position in degrees. (Remember this is only an approximation. Check your answer by looking up your latitude on a map.)

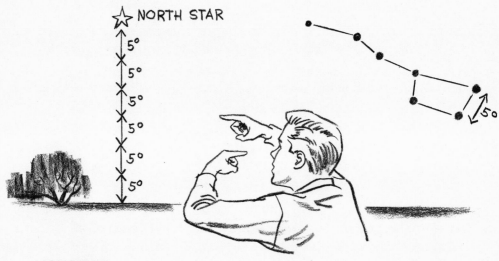

LONGITUDE

A ship's navigator may have found that his latitude is 40 degrees north. But in order to locate the ship's exact position another line running north and south that will intersect the latitude line is needed. The lines of longitude are used.

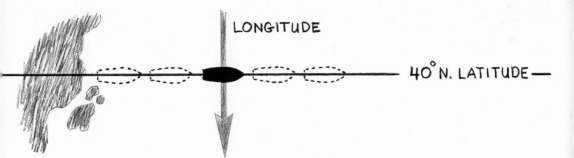

LONGITUDE

40° N. LATITUDE ——

Longitude lines are also called meridians, and are 15 degrees apart. The Prime Meridian is the zero line for degrees of longitude. It passes through Greenwich, England, and divides east and west longitude. The International Date Line is 180 degrees from Greenwich, or halfway around the globe.

For every 15 degrees, the clock is set back one hour if you are traveling west. A ship's navigator can determine his longitude by checking a clock set to Greenwich time. If it is 2 P.M. at the Prime Meridian when it is noon where he is, then his ship's position is 30 degrees west longitude. He is 30 degrees, or two hours, west of Greenwich, England.

TIME ZONES

There are 24 time zones around the world, each 15 degrees apart. The United States falls into four time zones, known as Eastern, Central, Mountain, and Pacific. When it is 11 A.M. in New York City, it is 8 A.M. in San Francisco. As a person travels from one time zone into another, his watch has to be set ahead or back, depending upon the direction in which he is traveling.

Latitude and longitude and time all involve mathematics and today, with our modern means of transportation, they are more important than ever. Safety of ships at sea depends on precise navigation. With jet planes traveling at 600 mph or more, positions must be figured accurately and quickly, and passengers must adjust to hours gained or lost with amazing rapidity.

MAGIC SQUARES

Long ago when numbers were still new to the people of the world, there were many men who experimented with them. They marveled at some of the things they could do with them and passed on this information to the people of their cities.

Numbers were regarded so highly that some of the people worshipped them and believed that they held the secret to life itself. The magic square was regarded by many as real magic. It was discovered by Emperor Yu in the year 2,200 B.C. Emperor Yu did not use our numbers because they were still to be invented. But the Lo-Shu (fig. 1), the first magic square, used symbols that can be represented by our numbers (fig. 2). If you add any row, column, or diagonal, it equals fifteen.

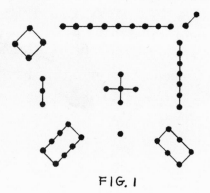

FIG. 1

4	9	2
3	5	7
8	1	6

FIG. 2

There is not much magic to this other kind of magic square (fig. 3). Yet if you add any row, column, or diagonal, it equals nine.

FIG. 3

3	3	3
3	3	3
3	3	3

ACTIVITY

Complete the magic square by inserting the proper numbers in the empty spaces. (answer on page 64)

16	3	2	13
5		11	8
	6	7	
4		14	

Symbols are a very important part of mathematics. They make it easier to write mathematical statements. You are already familiar with some mathematical symbols. Here are some which you may have used in solving problems:

+	add	**╱**	inch, second	**⌐**	division
—	subtract	**╱╱**	foot, minute	**△**	triangle
✕	multiply	**o**	degrees	**∠**	angle
÷	divide	**=**	equal to	**π**	pi (3.1416)

Scientists and mathematicians have many other symbols. Here are some that they use:

∠s	angles	**▱**	parallelogram	**—**	line segment
△s	triangles	**∞**	infinity	**⌒**	arc
‖	is parallel to	**≠**	not equal to	**>**	is greater than
<	is less than	**≅**	is congruent to	**≥**	is greater than or equal to
⊥	perpendicular	**⊙**	circle	**▭**	rectangle
∴	therefore	**≡**	is identical to	**∿**	is similar to

Some of the symbols have a long history. The equal sign we use today was the invention of Robert Recorde. He devised it in 1557. There are other ways of expressing an equality. Here are some that were used up to the 16th century:

$$\lceil \quad || \quad 2/2 \quad \supset\!\!\bigcirc$$

Write the following sentences using mathematical symbols. Example: The angles of the triangle are equal.

The ∠ s of the △ are = .

1. Angle one is equal to angle two.
2. The two triangles are congruent.
3. The arc of the circle is less than three inches.
4. The angles of the parallelogram equal three hundred and sixty degrees.

(answers on page 64)

Mathematics has served the sciences for many centuries. It has helped to make our lives more comfortable and enjoyable. Mathematics will always reign as "the queen of the sciences."

p. 10

303 = 𐎛𐎛𐎛|||

37 = ∩∩∩ ||||/|||

1,396 = 𓂀 𐎛𐎛𐎛 ∩∩∩ ∩∩∩ ||| / ∩∩∩ |||

32 = ∩∩∩ ||

21 = ∩∩ |

301 = 𐎛𐎛𐎛 |

41 = ∩∩ / ∩∩ |

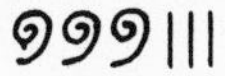

650 = 𐎛𐎛𐎛 ∩∩∩ / 𐎛𐎛𐎛 ∩∩

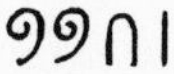

211 = 𐎛𐎛∩ |

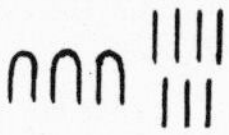

89 = ∩∩∩∩ ||||| / ∩∩∩∩ ||||

879 = 𐎛𐎛𐎛𐎛 ∩∩∩∩ ||||| / 𐎛𐎛𐎛𐎛 ∩∩∩ ||||

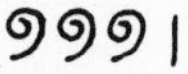

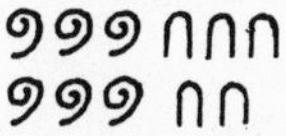

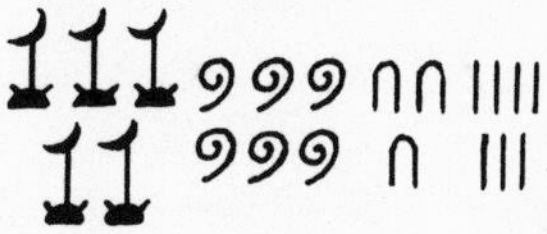

5,637 = 𓏢𓏢𓏢 𐎛𐎛𐎛 ∩∩ |||| / 𓏢𓏢 𐎛𐎛𐎛 ∩ |||

p. 14

67 = ⌐△Γ||

51 = ⌐|

425 = ΗΗΗΗ△△Γ

6 = Γ|

1,643 = Χ⌐Η△△△△|||

2,304 = ΧΧΗΗΗΗ||||

p. 15

LXVII = 67	MCMXIX = 1,919	MMCIV = 2,104
CLXII = 162	MCCCXXIII = 1,323	MCMLXXIV = 1,974
MCCXXII = 1,222	MMCMXVI = 2,916	

p. 16

964 = CMLXIV	1,976 = MCMLXXVI	1,241 = MCCXLI
1,872 = MCCMLXXII	2,306 = MMCCCVI	3,935 = MMMCMXXXV

p. 18

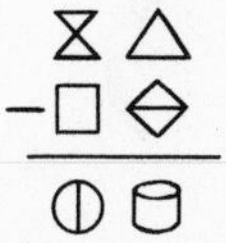

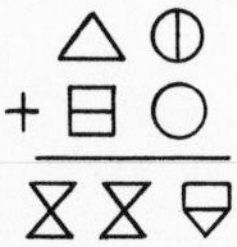

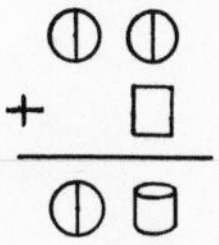

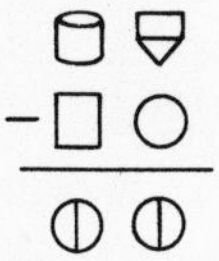

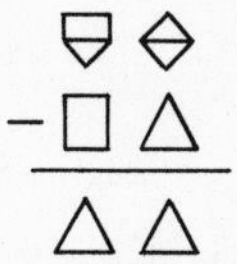

p. 26 1. 11:30 A.M.
2. 5 hours
3. 4:30 P.M.

p. 30 $462 = 4(10)^2 + 6(10) + 2$ $3,160 = 3(10)^3 + 10^2 + 6(10)$
$511 = 5(10)^2 + 1(10) + 1$ $3,156 = 3(10)^3 + 10^2 + 5(10) + 6$
$1,340 = 10^3 + 3(10)^2 + 4(10)$ $7,540 = 7(10)^3 + 5(10)^2 + 4(10)$

p. 36 10

p. 60

16	3	2	13
5	10	11	8
9	6	7	12
4	15	14	1

p. 62 1. ∠ 1 is = ∠ 2

2. The 2 △s are ≅

3. The ⌒ of the ⊙ < 3″

4. The ∠s of the ▱ = 360°

64